Prized Ponies

Narrative and Lyric Poetry

John D. Thompson

First Edition

Vista Press
West Des Moines, IA

Copyright 2000, John D. Thompson

All rights reserved: no part of this publication may be reproduced, stored in a retrieval system, or transmitted in any form or by any means, electronic, mechanical, photocopying, recording, or otherwise, without the prior written permission of John D. Thompson.

Manufactured in the United States of America
Library of Congress Catalogue Number: 00132651
ISBN: 0-615-11375-3
Cover Design: Karisa Runkel

Vista Press
West Des Moines, IA
515-225-0976, fax 515-265-3044
E-mail: jay_thompsonpoet@hotmail.com
Author can be reached by the above phone number or e-mail
For book sales, signings, and readings.

This book is dedicated to the people of my hometown, Lamoni, Iowa, and to the many roses in my life for whom I run.

Special thanks to Mike McCoy, Billie Evans, Dennis Howe, Donald J. Shields, Peggy Henson, Arthur Neis, Marian Clark, Vickie Anderson, Kent Vlautin, Clay Fogarty, John Fogarty, Val Russell, Todd Homan, Gene Shiro, Heather Steele, Linda DeBarthe, Janet Sylvester, Kathy Berdan, and my parents.

Thanks also to the faculty, staff, and students of Waukee, IA, Schools

A special tribute to the lives of Helen Hampton and Edie Shoemaker.

Table of Contents

THE STARTING GATE
Prized Ponies...11
Heartland Harlem...12
Bluegrass...13
Flow of Thought...14
Where No One Listens...15
Flower Dance...16
Dreams: A Tour of Duty...16
At War with Myself...17
The Decrescendo of the Balladeer...18
How and Y-Knots...19
Lady...Bull...20
Debt: A Rhyme Royal...21
Feeding the Seed...21
Showstopper...21
Diagnosis Dysthymia: A Gathering of Blues...22
Knee Bends...23
The Child...24
Ivy Leave...24
Mist of Grace...25
Aubade...25
Miracle Baby...26
Ayn Rand Is Hooked on Junk Food...26
Missing Child...27
Depersonalization...27

THE FIRST TURN
The Twenty-Five Cent Horse...31
The Mite Society...32
Summer Dress...33
Alumnus...34
A Horse of a Different Color...35
My Father's Will...36
A Marriage Proposal: Twelve Words, Twelve Syllables...36
Jonquil...37
Entrance Exams...38
Dream Dancer...39
Blood Bank...40
Breakdown and Afterglow...41

Spring and Release...42
Autumn and Fall...43
Happenstance...44
Watermarks...45
An Illiterate Kidnapee's Plea to a Major League Dugout...45
The Registrar's Arrangements...46
Invited Guest Absent...46
The Homesick Are Haunted...47
Civil Wear...48
Fatherly Advice...48
About a Horse...49
Stoops...50
Bereavement Ball...51
A Trio of Triolets on the Nuptial Theme...52
THE BACKSTRETCH with a quarter mile of quatrains
Cavalcade...55
Easy to Envy...56
No Donuts in Sedona...56
Branwell...57
Solipsism...58
Sea Son...58
Ghost Writer...59
Grave Expectations...59
Imagine That...60
Mental...60
Venal...61
What Is Wrong With This Picture...61
Honor Graduates at a Class Reunion...62
Lies to Launch a Thousand Ships...63
Blatant Nuances of Literal Irony...63
Mediums of Exchange...64
The Eight-Foot Bride of Harvard Sqaure...64
Rocket Science...65
Affair Matters...65
Fronts... 66
Jockey Shorts...66
The Rime of Rhyme...66
Unarmed...67
Unrequited...68
Cooking Geese...69
Dali Counterclockwise...69
Rational People...69
Couch Trip...70
Pricks and Roses...70

Revisions...71
Holiday Vignettes...71
The ABC's de Ecstasy...72
THE FINAL TURN
Whippers and Trainers...75
Oz Has Spoken...76
The Meadow Grounds...76
Passions of Pompeii...77
Wilde-ing...78
Old Early...78
Sonnet.com...79
Low Noon...79
From Fiction to Folly...80
Lying, Still...81
Equinox Evolving...81
Card Games...82
The Silences Know...83
Mirabilia: Wonder of Lost Love...84
A Civilized Man...85
Cantina Sestina...86
French Lick...87
Holesome...88
Church Mouse...90
Wild Desert Rose...91
THE FINISH LINE
Breaker...95
Flying Days...96
Dead to the World and Wide Awake...97
Philosophical Residue...98
Strung Instrument...99
Tolled Life...100
Bread and Circus...101
The Anti-Paradise...102
Everybody's Angel...103
Winter: An 8 by 8 Portrait...104
Earning Marks...104
Horsemen from Heaven...105
Babes in Basins...106
The Haze of Disaster...107
Between Good Friday and Easter Sunday...108
Century Falling...109
The Scent of Your Ghost...110
The Going Rate...111

Foreword

    At the end of the century, America had the privilege of discovering a new poet whose old-fashioned writing style seems remarkably fresh. John D. Thompson's *Prized Ponies*, the follow-up book of poetry to his award-winning, critically acclaimed *Hidden Voices From Hush-a-bye Rooms*, takes the reader on a chronological ride along an oval track, an extended metaphor for the journey of the human race itself.

    Thompson's writing style is a unique hybrid of Poe, Dickinson, Auden, and Yeats as well as some of the cleverest satirists of our time; but his work is no imitation. His verses are original, thoughtful inventions that will touch even the most hardened heart and challenge the most intellectual of minds. From Harvard to Hollywood, he is becoming one of the most celebrated poets of the new millennium.

    For those of you who think rhyme is a thing of the past, think again. Like Sandburg, John D. Thompson is a poet for the masses who can rub shoulders with the modernists anytime; but Thompson seems to be in his own time.

Dr. Donald J. Shields
Professor Emeritus
Indiana State University

THE STARTING GATE:

Whoever you are
Whatever you're worth
The race, little star
Begins at birth

PRIZED PONIES

Blue ribbons and badges
Sewn on a sash
Throw youth confetti
Ignore the trash
Prize your young ponies
With all that fine stuff
The real world strips
Those decorated delusions
In time well enough

Heartland Harlem

Beyond the supine fields of corn glazed with gold
Above the verdant valley where the determined river rolls
A people plant their lives on patches of pasture
Rumored to be green

We break out the bottles of bluegrass moonshine
Then party down under the poverty-level line
A prayer in church to ward off the devil and drought
We live from within, though we live without

Bale the hay, corral the cattle, spin the endless yarn
Gather for potluck in the ghetto disguised with fences and barns
Upon an early-morning porch,
This week's widow greets the day and weeps
As a rise-and-shine rooster wakes the dreaming from their sleep

I'm going back to Heartland Harlem
I'm going back to Heartland Harlem

A gasoline pump bursts with a broken gauge
A waitress busts her butt for less than minimum wage
A band of boys ropin' for the midsummer rodeo
Billy Ray bumps Charley's pride off the radio

A county-fair queen perched alongside a prize-winning steer
A smile of satisfaction stretching from hoof to ear
A family feud over a will bequeathed
A fortune found in the formula of crystal meth

The widow hides her stash between a quilt and shawl
Drug dealing done in flannel shirts and bib overalls
The local color is all tongue-and-cheek
A place for the struggling strong and the mighty meek

I'm going back to Heartland Harlem
I'm going back to Heartland Harlem

BLUEGRASS
For brother

Above bluegrass fields
The projectile of a pigskin brings the autumn on
Last summer's color-faded clothes priced and strung
Along jaded bluegrass lawns
The fall equinox presents its annual divide
As sun and moon quietly collide

Blades of bluegrass against my homesick heel
Blades of bluegrass against my homesick heel
A brush with boyhood I fight not to feel
Blades of bluegrass against my homesick heel

Was it Art of Paul who wrote *My Little Town*
That duet of brilliant boys should lay their pens of pathos down
I strive in vain to listen with nonchalance
I strive in vain to become a man of renaissance

The bluegrass hedges its nostalgic greenery
My mind becomes ether diffusing into the porous scenery
A mental high of anguish caused by a domain displaced
Losing one's mind is another phrase for losing one's place

These intake eyes invent alternative imagery
The bluegrass raked, bundled, and tied
A package refused upon delivery
Still, the song of bluegrass serenades me until death
As bluegrass music pipes into my last breath

Flow of Thought

I sink of sail along the current, Perception
Mystic brew, clear-blue, color of the sea
Filling my mind-matter with the hope of liquidity
Downwind of absolute lucidity
The promise of a great river passing
Through the backyard of my foreground
So roll, roll on, Perception
Spill your invisible wine
Pour the overflow into my thirsting eyes anew
And bathe in the cleansing of your freshest waters
Your stream of consciousness whirling into my waking conscience

WHERE NO ONE LISTENS
For Billie Evans

Listen, unheard one, where no one listens
A silence cries out to you
Like an orphan knowing precisely
Whom she wants as a parent

Listen, unknown author, where no one listens
In branches, poetic birds exchange plot-verses thick
With oxymorons, cumbersomely coordinated
Lines end-stopped and enjambed
Songs just short of epic proportion
For human language to complete

Listen, unpushed pen, where no one listens
The hush-hushes of lovers become privy to you
Their adhesive invectives and discordant confessions
Spilling and channeling
Into the rivers of your reference pool

Listen, unspoken voice, where no one listens
Life beats a wealth of wisdom on the drum of your ear
Tongues below your knee speak in rhyme and are free for a time
Tongues above your knee speak in free verse yet seem bound

Listen, unsung scribe, where no one listens
I apologize in some otherworldly whisper
For judging you without knowing you
For knowing you without judging you
For failing to recognize the necessary balance
Of your struggle from the unheard to the unsung

We might move beyond acquaintance, you and I
And rise into empathy
As we discover together the beauty above the noise
The mystery below its silence

FLOWER DANCE

In early spring
The leaves of hardwoods play musical chords
As blue pasqueflowers begin the Beguine

God's green grows deep
Bloodroot and violets tango in time
To the pulsing tempo of April rain

Road-ditch rhythm
Wild roses invoke summer's sweeping waltz
Purple phlox, prairie lilies ballet-bloom

Fall's curtain lifts
Goldenrods debut autumn's cotillion
A duet minuet of earth and moon

DREAMS: A TOUR OF DUTY

Wandering dreams
Meandering their lazy-work sway
Through the labyrinth of unconsciousness
Knocking on moonlit doors closed for the day
Opening ports to the flood of your thirst

Fondling dreams
Fingering their bone-probing way
Through the deepest crevices of sleep-thought
Waking old partners to act out a play
Reviving dramas long ago rehearsed

AT WAR WITH MYSELF

A mother's labor day is one of unrest
Until the soldier of triumph is laid upon her breast
Training at home, she arms him with guerrilla gear
Then marches the moppet off to battle among his peers

The wounded plebe returns home from his first tour of duty
Mother, I am at war with myself
I have no greater enemy than myself
I color in combat against myself
The teacher reports that indeed I am strong and I am smart
Yet I am weak in so many intangible parts
I am the destroyer of innocent children's dreams
A prodigious strategist mastering selfish schemes
Impressively, I tip the IQ scales
Yet the human inside miserably fails

The woman kneels to him and whispers
Son, your soul is a ring of fire
Lord have mercy on those who want what we desire

The embattled boy turns from her and silently surrenders
I am a child of ancestral anger
Born to duel domestic danger
I am the repository for my mother's uncried tears
Saltwater suffering known to me since my pre-natal year
I am at war with myself

THE DECRESCENDO OF THE BALLADEER

A crescendo heard throughout the land
A baritone bellows; his stature grand
Hands whisk wildly in the concert hall
As eyes conceal visions of an inevitable fall

For when the music mutes and the song subsides
The critics cross their *t*'s and dot their *i*'s
The balladeer bows perfunctorily, a reluctant choice
Like an arid Ariel stripped of her maritime voice

Yet there will be time to regather and recover
Time to lose your passion and settle for another
A time to delight and a time to divorce
When voices pledged before God grow weary and hoarse

And the balladeer now shining shoes
Tips his hat and bids bless you's
As trenchcoats pass and turn down darkened streets
Toward private parlors where capitalist conformity meets

And those hands who but applaud in concert halls
Clasp the coinage for labors common and small
Gold for greed, silver for schemes
A coin in the shoeshiner's cup to invest in a dream

For a time to rejuvenate and rebuild
A time to join the privileged guild
Time to push papers, time to punch clocks
A time to tuck away the balladeer's frocks

So you, Choirboy, with lungs and longings three octaves deep
With mind stuck on stage as you wake and sleep
You will sacrifice soul and forgo friend, though
Your song will best be known for its decrescendo

HOW AND Y-KNOTS
For DH

Rarely but often, I am proud of what I regret
I harden to soften the memories I forget
I seem so straight, I appear queer
The one furthest from my heart is the only one I hold near

To see clearly, I close my eyes
I shed my skins to reveal my best disguise
My mind is mute, and I am wise
I speak the truth to tell my lies

Steady in my stance never to shake with the hands of time
I am most innocent of all my confessed crimes
A player so ahead in the game that I am late
My heaven has frozen over, and Hell will have to wait

I take a leap of mistrust
A back step of faith
I enter through exits
To confuse those merely standing at the gate

LADY...BULL

Buffalo girl
Tending to men at the bar
Dream Horsey took you for a ride
(though not very far)

Buffalo girl
Crying on your birthday saddle
Too stubborn to sink, too jaded to swim
So strap on the latest beau and paddle

Buffalo girl
Serving up buffalo chips
With every sway in your gait
Every lick of your lips

Buffalo girl
Star-fucking in a starless pasture
Lying on your back, lying to yourself
Not knowing what you are after

DEBT: A RHYME ROYAL

Our Lady of Spain has moved to San Juan
Her kingdom resumed in tenement halls
Crowned jewels scattered along sandlot lawns
A fortune forfeited in market malls
No minions to master during the fall
Clipping her credit cards, making a mess
Keeping her creditors under her dress

FEEDING THE SEED

Oval of innocence, a seed so small
Embedded in the soil of the mind
Time nurtures and waters you
The liquid of life penetrates your naïve shell
As you open and flower to the colorless world

Father Sun and Mother Moon flatter and feed you
As you unfold and bloom
The bowing petals of inferior foliage applaud your appearance
A fury of wind inflates your efforts
The weeds beneath you attempt in futility
To drain your expanding beauty

Still, you rise in your shrine
A garden of one
A solo bouquet

Ever
Growing
Oval of existence

Surrounded like a royal castle
By a mote of minions who dream of you by night
Servants who wake to tend to you by day

SHOWSTOPPER

As I take cue on the stage of life, the air like a curtain parts
To unveil another step through life
And as I sleep, though the drapery departs
A blanket rewards me with good-night

I rise each day to the Director's call
A performance with purpose and cause
Although the audience is but the air itself
My clapping feet provide applause

DIAGNOSIS DYSTHYMIA: A GATHERING OF BLUES

A stetson unable to wrangle his stable of horses
A gardening glove raising a field of fools
A field of fools

A bottle of peach brandy cornered in the pantry
Drink the disappointment down
Secrets stored safely from the voices of this talkabout town
He is the commander of his own chaos
And she has enabled him a time or two

Together, their duet will coo
A gathering of blues

Big brother bounces on Daddy's directive knee
That tiger tyke gave every effort to please
But you have not been born, son
Until you are run out of Dodge
So who are the Marshall and his sidekick ever to judge me
So provincially

Baby blue
Glory boy blue
Rebel blue
A gathering of blues

Unhand the untrammeled ones
I beseech you
Let me take them with me
And together we will be
The lost children for eternity

The stones of a sepulcher stirring in my mind
A sentimental, sopping stew
No medicine man alive can cure me of these ills
So, My Savior, if you do
My soul of indigo belongs to you

Baby blue
Glory boy blue
Rebel blue
A gathering of blues

DIAGNOSIS DYSTHYMIA: A GATHERING OF BLUES

Safeguard the unseasoned ones
I relinquish
The malleable are best molded by you
No one does the dance of dysthymia
No one loves the language of languor
Like I do

As sea waves roll over
Another day washes ashore, anew
The maudlin haze cloaks the morning dew
The day breaks in a familiar hue
A familiar hue

KNEE BENDS

Child,
Your bended knee a kiss
Upon the soil of eroding tears
Pray for better days

Your ancient youth, a sentinel to the pain
Your knowing naivete, a warden of the weather watch
As winter crawls upon the back
Of your summer street

A broken lamplight, a sidewalk unattended
Except for blue-collar feet
Scampering for wages
Or idle toes killing time

Idle hands, you are told, are the devil's work
As if Satan had no other objective
Than to plague your family
With a swarm of lassitude

Food stamps arranged like a game of solitaire
Upon a coffee table stained with failure
Its worn legs bowing to the sofa older than your old man
A sofa soiled each day by the dirt on your knee

THE CHILD

The child refused to participate
The child refused to grow
The child could only rue and hate
What adulthood had to bestow

The child surrendered to his inner child
The child refused to learn
The child became weak and wild
When Time took its uncontrollable turn

The child is so unable
The child, quite unaware
The child spat upon life's maturing oak table
Then collapsed in his rocking chair

The child ignored his purpose
So Life took him by the horns
To lay him beside a bed of roses
Among a thatch of thorns

IVY LEAVE

The final score
SAT-1280
FEAR-a perfect 1600
Numbers don't lie
So much for the old college try
A farewell to Ivy, farewell

MIST OF GRACE
For Marian and her Oregon family

Grace walks with me to greet each day
She moves in quiet precision
Grace offers light to suggest the way
She whispers my life's mission

Grace lends a hand when I must fall
She reminds me to stand up straight
Grace goes in peace when duty calls
She is early when Hope is late

AUBADE
Haiku

Step lightly, miner
Lantern rising from the night
Gleaming cache of gold

MIRACLE BABY

Every other minute
A miracle baby is born
Mama's mincemeat of pride and joy
Papa's bread and butter

The first demographic
Male or female
Lies between the child's legs
The second demographic
Date of birth
A lie to be told in time
The former you hide by the seat of your pants
The latter you hide on the tip of your tongue

And what do we do when the miracle baby
Enters a less than miraculous world
The womb has no return policy
Papa has bolted the door
Until you come home a success

Our miracle baby becomes fresh meat to the market
He raises bread to supplement loafing loaves at home
And churns butter on the street to replace rancid cream

AYN RAND IS HOOKED ON JUNK FOOD

No human being produces more than he or she consumes
The filthy things gather coins and titles
Along the hierarchy of Humanhood
Symbols tangible and abstract
To convince their mammalian minds
Into believing otherwise
Nothing feeds on your resume like your ego
So keep buying, keep breathing,
Keep existing, keep eating
Stay focused, my friendly user, on the sway of the clock
Master symbol of your tentative time
The hypnosis will not work on you
Objectivism, my ass
The Fountainhead foams at the mouth
Atlas Shrugged because he does not know and does not care

MISSING CHILD

A missing child's mug shot joins me for breakfast
Milk-carton tears fall into my bowl
His stats and description look familiar
We possess the same heart and soul
The world swallows us whole

My father's voice could never reach me
I longed to sing a different song
My brother's noise could only teach me
Everything I did right he deemed wrong
Where do I belong

A vagrant on the road to nowhere
My protruding thumb asks the horizon for a ride
Someday, somehow I'll end up somewhere
With my hometown cheering at my side

The northern lights will redirect me
To the heights I have never known
My mother's prayers will resurrect me
To the womb of her hidden home
Where I live alone

DEPERSONALIZATION

I stood back to watch the world pass me by
A spectator of life's cruel parade
My eyes wide open, still nothing descried
Darkness on light, each shadow a tempting charade

Like Rodin's *The Thinker* marbled on post
My state of mind a naked-dream nightmare
Familiar faces as phantoms and hazy ghosts
I pray for a time when the fog might clear

A man-child body in juvenile jail
The warden myself denying release
As free men learned to win, this captive failed
The hope of tomorrow resting in peace

THE FIRST TURN:

Veer to the left
To stay on course
A former filly is bereft
As her babe becomes a horse

The Twenty-Five Cent Horse

Place your quarter-bits on the quarter horse
Watch him run about the town
Dangle carats before him when the stakes are high
Kick his fallen flank when he is down

Wager your silver on QuickSilver
When his shimmer fades, turn on him like a dime
The neigh-sayers shoot horses, don't they
When those manes of magic pass their prime

THE MITE SOCIETY*

The full-moon bases of the teacups formed concentric circles
Upon doilies of latter-day lace
As the ladies of leisured labor assumed seated position
Each square of cloth, each dance of thread in proper place
Informal salutations sufficed for a roll call
Before graying foreheads bowed for a pause of piety
As the patchwork of another day comes undone
I wonder what is needling the Mite Society

The oldest pair of palms spoke of her deceased husband
An act of heroism she struggled to understand
If the same fate were to befall her enlisted grandson
She vowed to tear NATO down with her bare hands
A set of naïve nails rattled about an unexpected pregnancy
A rumor spawning from the local junior high
An aerial attack by an artillery of shaming index fingers
Suggested the rumor was merely a lie

Supine and swaddled in a quilt of comfort
A blanket bestowed on my graduation day
As I woke to wrestle with the arms of morning
A call from the cloth caused momentary delay
Daughters of the woven dream
  Scions of Zion's scheme
    Sisters of the seam cloaking their naked young

God bless and keep you
Throughout the travels and travails of your life
As we wrap you in wisdom, our native son

*The Mite Society is a quilting club in Lamoni, Iowa.
It is believed to be the oldest of its kind in America.

SUMMER DRESS
For Janet Sylvester

You hang and wait for the wind
To blow sweet-somethings
Underneath your breathing hem
Like a courting beau

A playful whisper of seduction
Amidst the humid holler of June
A summer dress swims in the deepest drought
It moves the stillborn air

At every turn, a dancer
At every pause, a dream
A summer dress speaks of lipstick sunsets
And the rise of an August moon

Easy on the eye, easier on the flesh
A summer dress flows to the soft cadence of the season
Wafting and waltzing to the serenade of a whistle
Lightly lifting to accept the call

A summer dress, any man will confess
Succumbs him until the rise of Fall

Alumnus

You broke bones on this field
When your marrow was young
As did your rivals, as did your son

You stand on the sidelines with the retired rest
An alumnus wondering where his past has scattered
As if it meant something, as if it mattered

You taught at this book-barrack
Some of the enlisted thought you were smart
A soldier of science, the methodical art

I did not want to catch your pitch, alumnus
I chose not to bite your bits of gold and blue
I wanted nothing of this graced land's feast
I wanted no part of you

I will not stoop for your heights, alumnus
I would rather fail or crawl
Than to stand among the self-proclaimed mighty
Rehashing the haunts in your alumni hall

A Horse of a Different Color
For Todd

I came across a horse of a different color
A hoof of another hue
I took him by the bridle to become my groom

I met a horse covered in black
A steed mounted for the next attack
A shadow-soldier of unknown valor

Oh, before my eyes I could see
A carousel of outcasts revolving around me
The argyled, the damasked
A color and style for every taste and task
A trotting tapestry
With signatures by artists unknown

Beyond hill, mound, dell, and glade
An escapade in every shade
A prismatic parade toward a hidden home

I came across a horse of a different color
The sight raised me beyond the blue
I took him by the bridle to become my groom

My Father's Will

My father's will is thrust on me
His tone of voice
A muscled arm
Those papers of legality

So I am seen yet seldom heard
Except when replying to his word
When I inherit my father's will
I vow no child's blood to spill

Whether by the will of promised fortune
Or the will of one man's brute force
I will not carry his legacy
Further down this tortuous course

A Marriage Proposal: Twelve Words, Twelve Syllables

Here is a *damn*—
If you take it,
I will give it

JONQUIL

I bloom the flower jonquil
A petaled Narcissus
Thirsting for a pool of reflection
Hungering to feast upon a mirror
To return to me my cornucopia of beauty
My ambrosia of natural gifts

I am white for the gods have made me pure
Among your tainted madness
I am yellow for the sun rises and touches only me
Among your lilies of The Field
In down time, absent of light
I pine for my moments of glory
My visibility taken briefly by the hands of darkness
Captured selfishly by the charred fingers of dusk

But Night is no foe for my rising ambition
My resplendent determination to appear
In the spotlight of morning
As I present myself to the applauding eyes of the world
I, an Evita on the green

How I missed you at midnight, my voyeurs
Yet weep no more your tears of moonlit dew
I bloom the flower Jonquil
At the peak of my season
And my season is never out of style

ENTRANCE EXAMS

Girl to woman
You passed the test
Man to boy
I passed my peak
To conceal the scars
Of this wrongful rite of passage
I turn the other cheek

Like a coin, we have flipped the rules
Men are from Venus; Women are from Mars
Our love could no longer dwell on this planet
So we cast its lifeless legs to the stars

I should have kicked and screamed
To bring your hasty wedding to a hault
I had nothing left to lose
The insolvency of our souls had been my fault
If I had killed all of the competition around me
You would have been mine by default

I wish were studying for those college entrance exams again
We were weekend lovers
The rest of the world rested in someone else's hands
Our parents' limited economy
Supplied us with anything we would demand

John Lennon up and died on us
Can you imagine that
We struggled to learn the struggles between
Israel and Yasir Arafat
Stitching alligators on generic shirts
Gave us the illusion
We knew where preppy-chic was at
Now our lean abdomens and bustling breasts
All have gone fat or flat

ENTRANCE EXAMS

When your chords began to sing for another man
My heart became an organ playing
Its own rendition of the blues
My history of using you taught me only
How it feels like being used
I searched for alternatives to my wicked ways
You would not give me the time to choose

I wish we were studying for those entrance exams again
We were weekend lovers
Your heart rested in my open hand
Like an hourglass broken at the base
Our love became a pile of sifted sand

DREAM DANCER

I, the dancer in your dreams
Tango-tripping over your fantasies
My stilettos scarring the nocturnal skin of slumber
The cadence tapping against your cranium
Disturbing the mellowing song of your sleep
I dip into your REM as you keep optic rhythm
To the groove of my entry
Two-stepping on your resting heartbeat, your fading brain beat
I ignore the supplications of nightmares to enlist on my dance card
Tonight, it is just you and I
Your medulla oblongata breathes *hakuna matata*
No worries

BLOOD BANK

The first time
I thought it was the last time
I would ever see you again
The pulse of living escaped me
Siphoning envious blood and hapless rhythm
From my dying heart
And giving you more hemoglobin than your share

You said I did not know you anymore
You announced you had gone to Graceland to play grown-up
So that you could escape the nightmare of your childhood
You told me you needed my extra plasma for your new lover
He had grown anemic with all of your love-making
Your vertical demands, your horizontal whispers

I, of course, heard none of this
I focused on the deafening departure just minutes away
The weakening power of goodbye
Apathetic to your words
Unreceptive to those superficial gestures that make a girl a girl
Catatonic to your dreams
I became clotted stones
Wanting to break you and bleed you
The blood spilling over the fields of our folly
So no one could sip from this tainted red river
Sparing your new suitor of my vengeful poison
Yet depriving him of the passionate nutrients
Necessary and sufficient to quell your parasitic thirst

Breakdown and Afterglow

To come here, to these Ivy walls, was your biggest mistake
The streets were too crowded
The books were voluminous
The syllabus was too swift
The competitive minds were too conniving
The curriculum was too corrosive on your mind of fresh eggs

--But dropping out and breaking down, Plebe,
Need not go hand-in-hand
You could have gone home to smell the wild roses
Instead of wreaking with the stench of failure
You could have gone home
To write the memoirs of your precocious childhood
Instead of penning sham suicide notes

--But high drama became you, so you became depressed
You began to worship the fallen idols
Sylvia Plath, Janis Joplin, bell jars, and pearls
Tragic flaws with tragic endings
All that brilliance so misunderstood

Your hometown's tummy giggled
As the feathers of your folly tickled their fancies
Before the hands of your Christian community tarred those feathers
The Germans called this term *Schadenfreude*—
Others rise and your demise
And a *kraut* term was appropriate for you
A star pink behind the ears
It was your turn for the gas chamber
Glowing among the noxious smell of laughter

If you stick around this chamber long enough, son
The density of gas turns out the lights
If you stick around long enough
Afterglow

SPRING AND RELEASE

The diva winds of December bow at encore and then are gone
The ides of March arise to sing
At the turn of Nature's tune, I, too, will be gone
My heart is released in Spring

To unbind my ties
To loosen my sails
To walk through the mists of April
Searching for fairy tales

The diva winds of December bow at encore and then are gone
My heart is released in Spring

Fret not, my love
In winter I will love you
We will rekindle what warmth the logs on the fire bring
Don't cry, my love
The winters love you
My heart is released in spring

Darken the lantern light and close your eyes
My fading footprints whisper goodbye
The diva winds of December bow at encore and then are gone
My heart is released in spring

Autumn and Fall

Give him everything in May
So he will appreciate nothing in September
Bury your anathemas for him
Deep inside your good-day biddings
Carry your curse-buckets to the wall of secret wishes

And when he leaves his rooted territory of familiarity
His Towering Tree of Childhood looming much taller than yours
Rake him when he returns a fallen leaf
Blowing from doorstep to doorstep
Begging for a kind word

Gather his withered being and place him among your collection
Of tumbled, troubled souls
Arrange them in an oval upon your dining-room table
The leafed oval, a centerpiece
For your Thanksgiving feast
Dig in

HAPPENSTANCE

Things happen in our lives
Naughty, intrusive things
They happen in our lives

An unwelcome house guest
An unwanted stranger
They happen in our lives

A busted, rusted pipe
During a sleepless, weeping night
They happen in our lives

A reluctant goodbye
A required hello
They happen in our lives

The close of a show
The final episode
They happen in our lives

A runaway daughter
A discontented spouse
They happen in our lives

Ours is not to predict
Just where, when, or why
They happen in our lives

WATERMARKS

The dint of failure on your forehead
Like droplets of water
Single-file falling
An implication cascading into a confessional pool

I
am
the
bane
water
fallen
failure
collects

AN ILLITERATE KIDNAPEE'S PLEA TO A MAJOR LEAGUE DUGOUT
(Each letter of the alphabet used only once)

IMNTRBL
QK
FECH, GYZ

PS DU JOX WAV

Translation: I am in trouble. Quick! Fetch, Guys! P.S. Do jocks wave?

The Registrar's Arrangements

The sultan of scheduling, you offer the courses
Like Moses parting the sea
Your machinations separate the Christians
From the meandering lost

You give your son's hand
To the college's light of life scholar
She graduates him with honors
Then he is gone

You came in with Nixon and went out with Reagan
Though men of that stature you never knew
Still, your alma mater built you a haven
For collecting the tears of freshmen and tuition dues

So this is my tribute to the registrar's arrangements
Maneuvers checkered in gold and blue
You taught me well, my instructor of nothing
Those who cannot, register those who can do

Invited Guest Absent

Where was I at this wedding
An invited guest absent
A lined shadow on the groom
A fallen flower girl
Shown the exit by broom

The whisper of voices
Ricocheting the rumors
From front to back pew
Among something old, something new, something borrowed
I was something blue

INVITED GUEST ABSENT

Where was I at this wedding
When man and wife were presented
As the resounding congregation approved
In the exhaust of a runaway honeymoon four-wheeler   stood I
The boyfriend once removed

I was the stirring silence when the minister asked
If anyone present here...
Well, you know the rest
I was the broken heart beating
Upon your breast

THE HOMESICK ARE HAUNTED
For Vickie

A guinea pig on scholarship through the labyrinth of college
A tow-legged tester, tattered and taunted
By a spirit of pathos prolonging his past
The homesick are not sick; the homesick are haunted

The day an innocent bystander
As tomorrow prepares to bewitch and bereave him
His wounded anticipation reaches for the palliative past
Those documented days no longer need him

Shackled by change, the haunting hoorays of hometown he hears
When childhood crowned him the champion of cheers
Yet the crowds are capricious, the faithful seldom follow
When defeat thrusts its dosage down your throat
Demanding for your turn to swallow

A candidate for a commencement of confusion
He meanders through a melancholy milieu
A freshman foraging through a forest of papers
A term paper, peace of mind long overdue

A date with depression, anxiety clinical
A spectator of sorts as his life parades past its pinnacle
An orientation of angst, assimilation unwanted
The homesick are not sick; the homesick are haunted

Civil Wear

War, you see, is a matter of civility
I in best-man blue, my brother in gentleman gray
We draw our swords toward
The dawn of a dying day

Blood spills like uneasy wine
Below the belted Mason-Dixon Line
The bellicose become comatose
We kindred folk have grown too close

Field hands spar for cotton; sons stab for slaves
Privileged pens lay down the laws then lay our graves
War numbs the heart and succumbs the head
The birth of a nation decays until dead

Fatherly Advice

I was your gem; still, you told me man was loser
So I shed my shimmer and lost
You belted your desire for me to fall on my ass in college
I not only fell—I trumped you—I dropped out, daddy
And as I wandered aimlessly in the welcoming alleys
Of some strange and wonderful city
Going from man to man, father to brother
You took your pain to town
Weeping openly to your rural community
Your small audience of small ears
Pricked up like stalks of grain to hear the telecast of your dismay
You confided in them that you had done all that you could
Pouting, *The boy can do what the boy wants to do*
--Perhaps the boy just did—
How you wished your wallet could buy back time,
Refund your arrogance and purchase a place on earth safe for me
But I was secure among the insecure

FATHERLY ADVICE

You advised me to come home
To post those lofty GPA numbers like before
The ones that made you so proud yet envious
Turning your face from dollar-green to honest-green
Holding me captive in the paradox
Of papa's paradise of pride and envy
I, your amethyst; I, your emerald
Now a diamond in someone else's rough

ABOUT A HORSE

I saw a man about a horse
I saw a man about the town
I heard his silence speak volumes
Filling the white-noise streets with his sound

I saw a man take a tumble
Falling off the high horse named Pride
At least he dared to ride full mount
The horseman refused to ride saddle side

STOOPS

When he stoops to his knee
To engage you with a hole of gold
The cost of the finger-belt is on him
The cost of the gesture is on you

For the work of marriage becomes harder
When he plays on golf courses
While you stay in college courses
And every other couple around you divorces

You learn the hard way, little English majorette
Of the fetid irony hiding in your wedding dress
When he finally lifts your veil
To unmask his true motive
A mediocre fisher of men
Who needs a brilliant mermaid
To guide him through the high seas of academia
And you, desperately flapping to fetch his love
An Eve of the sea eager to take the bait

The hook he offers is a false tiara
My dethroned Duchess of his Dupedom
Yet the bouquet of flowers remains forever
All artificial things do
Now, you are on your knees
The left knee prays for his return
The right knee stoops to conquer

BEREAVEMENT BALL

My soul is a loosening thread
Clinging to a heart worn upon a torn sleeve
Love lingers until its dying destiny
When disenchanted hands apply for a license to leave
But if you still believe
In granting a discarded suitor one last reprieve

Meet me at the bereavement ball
A dance for charity, a favor in kind
As we move along the broken glass
Cinderella's slipper left behind
Meet me at the bereavement ball
For my bereavement bawl
A black boutonniere to offset your orchid of white
And we will go waltzing tonight

I foreshadow your searching eyes glancing across the dance hall
As those inviting ovals survey my way
The quiver of a lost romance crawls
A fool to take this chance
A certain victim of uncertain circumstance

The restless hour arrives
A note hanging like a noose from your forbidding door
The stiffened penmanship stipulates
I am to darken your doorstep no more
The caustic cursive makes me want to cry
The simple signature makes me want to die

Meet me at the bereavement ball
A dance for charity, a favor in kind
As we move along the broken glass
Cinderella's slipper left behind
Meet me at the bereavement ball
For my bereavement bawl
A black boutonniere to offset your orchid of white
And we will go waltzing tonight

A Trio of Triolets on the Nuptial Theme

Darling's Wedding Day

All else is white on darling's wedding day
My heart is dressed in dapper blue
The groom is tuxedoed in black; the pastor in gray
All else is white on darling's wedding day
A father gives my veiled Love away
The envy-green punch is spiked: I've had a few
All else is white on darling's wedding day
My heart is dressed in dapper blue

Disavow

This hand in marriage requires a ring
A circle of promises choking a finger
As passions fall, temptations spring
This hand in marriage requires a ring
Disbandment of union caused by a fling
A lovers' feud scheduled for Springer
This hand in marriage requires a ring
A circle of promises choking a finger

By the Powers Vested in Me

So I think not much of this institution
Where man and woman become entombed
I am obligated not by the Constitution
So I think not much of this institution
When betrothed lovers part what restitution
Can compensate a fallen bride and groom
So I think not much of this institution
Where man and woman become entombed

THE BACKSTRETCH:
WITH A QUARTER-MILE OF QUATRAINS

Some will stumble
Some will stride
Who is whipping you
On this ride

CAVALCADE

Cavalcade
Horses on parade
A fine line of thoroughbreds

Promenade
Like sure-footed gods
Greatness gallops in your heads

One in a million
To the vaulted pavilion
Claim the prize in the crowd you desire

Onward, cavalcade
To emerald pastures in shade
Oats of gold await when you retire

EASY TO ENVY

Beauty, you are easy to envy
Talent, so delicious to desire
Ambition causes mouths to water
Commitment sets it all afire

How these treasured traits rise in their hour of glory
Like concrete, these abstractions slab to the ground
        when they fall
Their memory is still easy to envy
These gifts bestowed to few and coveted by all

NO DONUTS IN SEDONA

No donuts in Sedona
The desert has no rolls today
The sewer lines are broken
Dust devils raise hell in the highway

No donuts in Sedona
Where earth tones paint landscape and sky
Deep-fat fryers are banned like bandit-outlaws
Let's give the bagel shop a try

BRANWELL

Branwell, our brother
If ever there be a black sheep
The lamb of darkness has to be you
Every painting you christen
An oil spill of failure
The pellicle spreading over your ocean of pain

Each sentence we Bronte sisters write, unquestioned delight
The Three of Us, a literary triumvirate
Immortalized in the pages of time
Your portrait of us seals our fantastic fate

We push the pens and don the pants
We dip for ink and dance the dance

You withdraw your brush and become a recluse
In the halls of Haworth Parsonage

Take solace, Branwell
We are the shepherds darning your dismay
With threads of wool from your heartbroken back

Weep not, Branwell
Mediocrity is not a crime
To err is human; To *Eyre,* divine

Solipsism

I have an eye to perceive the day
I feel your pain should you press my way
I prate opinions with precision
My tongue unable to slip beyond my vision

To my own self, I am referential
Even empathy is existential
So obligate me not to your objective
'Tis the world itself, selfish and subjective

Sea Son

Who goes there into the Sea Lion's lair
An empty vessel sailing aimlessly
Devoid of confidence, lacking coin
Entranced by your entreaty to join you
Through the aging, raging waters of these tired straits
A young wayfarer of the waves
So eager to pacify Old Neptune's complaints

Sail
On,
Sea Son

Ghost Writer

Ghost writer,
Phantom of the phrase
Shall I presume you dead
Each story rising from a sepulcher
A dying dictionary roaming around your head

Ghost writer,
Interred in a system
Where fairness makes no sense
The simple-minded buy your sentences
All at your expense

Ghost writer,
Credit escapes you like a spirit
Fleeing the soul of a hit-and-run victim at night
Your due accolades swept behind the moon
The haunts and taunts of your language igniting other light

Grave Expectations

I know not a flower to bloom in winter
Nothing rises from the snow
After the ascent of aster in autumn
Madam's chilling hand strikes a lethal blow

I know not a bloom to flower in summer
For my love failed to forget-me-not
I expect nothing from the rains of April
Except for the damp roots of my cellar-dreams to rot

Imagine That

I'm agog with image
Whatever image conveys
Image wets my lips
Image gives my loins a lay

Image is as image does
Image matters just because
Image has everything to hide
Image protects me from the inside

Image is short of depth
Image is lean of range
I will change my choice of image
Should image ever change

Mental

My mind is out of gas
I lament this burned-out brain
All cognition has gone incognito
I am out-of-touch; I've gone insane

My cerebrum is out of fissure-fuel
No functioning cells in sight
A mental faculty on labor strike
I am blinded by an absent light

VENAL

Ungratefulness, you serpent's fang
Venom to my vein
I gave to you my heart's content
A gesture most in vain

Thanklessness, you capricious wind
Fickle like a storm-tossed weather vane
Stirring clouds into my best intent
A glorious gale so vain

WHAT IS WRONG WITH THIS PICTURE

Leda had sex with a swan
    And gave birth to Greek civilization
My neighbor had sex with the Schwan's man
    And gave birth to a town scandal

Leda bore Helen of Troy
My neighbor bore in hell for her ploy

Helen of Troy's face launched a thousand ships
My neighbor's disgrace launched a thousand lips

A thousand ships began the Trojan War
A thousand lips sunk that Latex-Trojan whore

Honor Graduates at a Class Reunion
For Linda

As I diminish among the other wallflowers, she is more.
The extend of her gait surpasses my stationary shadow.
The belief in her faith reaches beyond my doubts.
The silver on her plate builds a castle of fortune
    around my castle of sand.

Another reunion of doctored resumes,
    and she must stand like a goddess,
    my high-school nemesis and academic rival,
Blowing a fortress of forbidding air between
    strategic sips of punch.
Her piercing eyes scan the room like a sniper
    looking for an assigned target.
Those erudite senses detect me as she positions
    her talons for approach and landing.
She looks at me concerned
    that I may be a candidate for Purgatory.
I gaze at her convinced that I am.
With an arch of her spine, she fires,
I hear you're going Ivy for some sort of writer's thing
—How'd you get in?

I want to extrapolate on some
    eclectic procedure of admission.
I want to tell her of a thousand rings of fire
    through which I had to pass.
I want to tell Miss Merit Scholar
    that she just finsihed a sentence with a preposition.
I shrug my shoulders and silently hoist a flag of surrender.

In my mind, I have made love to this woman
    countless times.
When she shines,
The beauty of her brilliance belongs upon a renowned
    painter's canvass.
When she stings,
The weapons of her brilliance belong upon the fields of war.

HONOR GRADUATES AT A CLASS REUNION

As I begin to speak, she winces and raises her lip,
Cracking her lipstick but not cracking a smile.
She looks at me with those smalltown eyes,
Those grapevine eyes that know all and tell all.
Her *summa cum laude* confidence
Pushes me further back into the bouquet of wallflowers.
Her Ph.D. persona trammels and trounces
My wounded petals into the gym floor.
As I disappear before her face of accomplishment,
I want to tell her I could not care less
                              --But she is more.

LIES TO LAUNCH A THOUSAND SHIPS
A cinquain

Exposed
Liar, confess
Your undressed undoings
Like soldiers from the Trojan horse
Tumble

BLATANT NUANCES OF LITERAL IRONY

I am convinced that I will never know
The meaning of the words cogent and ascertain
I am doubtful if I ever will learn dubious
I am filled to the brim with disbelief
That rife, replete, and incredulity
Are much too difficult to penetrate my brain
Like impervious

Mediums of Exchange
An Ottava Rima

He stole the moon to pawn it for a soul
She robbed the sun to switch it for a heart
The moon was full and it swallowed her whole
The sun was brittle tearing her apart
A thief of heaven must pay his due toll
A bandit of light deserves a rough start
The penalty high for the gifts from above
A baneful bliss for the bruises of love

The Eight-Foot Bride of Harvard Square

Her stance of silence and solitude veils the chaos and clutter
Of the bustling romp of Harvard Square
Like Dickens' Miss Havisham she awaits her prince
Though no passerby thinks she will find him there

She dares not to blink or to wince
The Eight-Foot Bride of Harvard Square
Rises above the beggars in the shops and citizens
        in the gutter
As suitors consider entering her carnival lair

To call her a mime insults her intention
The Eight-Foot Bride of Harvard Square
She offers a flower vase to The Paper Chase
Her betrothed beauty encased in every petal cast to the air

Somehow without person, but never out of place
Her thing is to bear witness, to bring peace, and to stare
I am frozen in awe of her invention
The Eight-Foot Bride of Harvard Square

ROCKET SCIENCE
For Val and Fred who never met

Open hand, grab a job
Life need not be rocket science
Where's your self-respect, Joe
Man, where's your self-reliance

Muscle in the sand, you oiled slob
Grunting and groaning like an angry tyrant
Lifting weights instead of freight
Working solely for your body's own appliance

Tater casserole on a couch, you lifeless knob
Watching the Jets fly by the Giants
Your mind becoming merely inner space
No place to study rocket science

AFFAIR MATTERS
Terza Rima

We meet upon the darkening hour
Among the shadows blue.
Love, shall I disclose my feelings now or

Wait for a sign from you.
I lock into your gaze at our table
You sense the passion, too.

I move to speak—but, alas, unable
To phrase the moment right.
My cowardly heart, yellow and sable,

Pounds with a depth of remarkable height.
Your budding smile a rose
Offered to me by way of candle light.

Tonight, you recline; my love for you grows.
Tomorrow you will rise—and so it goes.

FRONTS

A cold wind blew into our marriage
During our Arctic year
We shivered and worried
How best to survive the storm front

You fight fire with fire, you suggested
I winced at the cliché but agreed
So we erected a front to fend off the front

I wore my cloak convincingly
You draped yourself in true blue
No one in our suspicious circle ever suspected
Our home-fire's misconstrue

A cold wind blew into our marriage
My Ice Queen, you have nothing to fear
Just smile and act happy for the chilled children
The frozen never shed a tear

JOCKEY SHORTS

Silks on a saddle
A centaur of sorts
Whipping a log of leather

Loser by a length
Nosed by pinch
If only horses had feathers

THE RIME OF RHYME

Rhyme is dead; rhyme is frost
Free verse ices rhyme on the cross
Rhyme is frozen beneath the snows of Amherst
An art of antiquity to the fashion of free verse

UNARMED

Unarmed
I clench the padded handle of the final piece of luggage
The last vestige of your belongings to the front curb
I deserve these white knuckles
    forming a fist around my grip
I am entitled to this bulge of blue veins flowing
    like flooded rivers upon my wrist

The palm of my other hand nests atop our daughter's head
I cannot tell my little bird goodbye
So I lightly stroke the fall of her single braid
I feel the journey about to embark for her and you, Clarisse
My fleeting, fleeing daughter and wife
As both of you make an exodus from our unholy home
I feel the distance and solitude of all three of us
In this discovery of loss

My hand travels down our daughter's queue
Tight in its perfection, tight enough to hold secrets, I pray
I feel alone standing between you—
Though I have stood between you for years
Now, you have united, mother and daughter
Departing in the driveway
Space and time
All coherency about to escape in a runaway Mazda
The first farewell perhaps for the last time

And you, Clarisse, have become an enigma of change
Your face no longer a face but a head of stone
Upon which my waves of tears break and fall silently
When that face was mere flesh, I could break it, bleed it
--But no more
I can feel the freedom about to break wide open for you

With our bruises healed,
You approach me with a pen and paper
Our battling minds have forgotten
    to sign the terms of visitation

UNARMED

You hand the document to me, our skin careful not to touch
The paper is filled with stipulations, can's and cannot's
You offer me the Bic and say,
*It could be worse, Bill. You should be in shackles. Being a
father has kept you free.*

The pen possesses a slender sharpness,
        a dart-like verticality and firmness
Finally, something solid to hold
        during this evaporating day
I have a weapon to wield; I am armed
I wanted to do it—lose my mind—rage, stab, inflict
The braid holds me back like a rein
I sign with John Hancock flare,
        a peacock feigning pride of his failure
This could-be felon surrenders the goods over to you
I can stand no longer, I can stand it no longer
I walk to the house surrounded by invisible bars
        and phantom guards
Sit on the plush couch which feels like a cot and start
        serving time

UNREQUITED

He spoke in trimeter
I believed not a word
He vowed that he'd bleed her
His threats airborne, absurd

He wondered who'd need her
His love, free as a bird
Should fate at last leave her
Ground to the earth unheard

COOKING GEESE

Insanity is a squirt of lemon
On a paper cut
A solution of salt upon an abrasion
The sting of the sodium chloride never letting up

The crazy ones rotisserie in society's juices
Until they are so shaken or stirred
Even if their simmering senses could recognize the abuses
The seasonings hold them in silence,
        and the cooking geese are cured

DALI COUNTERCLOCKWISE

Ball and chain around the wrist
I am a slave to time
Ever watchful, every minute
Precious Time, include me in it

Peerlessly prompt to every tryst
Soft watches turn to slime
To procrastinate is to dissipate
Dali's folly a mocking tribute to the late

RATIONAL PEOPLE

Rational people, step aside
You are too tall to ride this ride
Rational people, shut your mouths
Turn your tails and head due south

Rational people, your logic is too hazy
To those of us clearly crazy
Rational people, such an ordinary find
Take our advice, lose your minds

Couch Trip

He spoke but could not read
She read but could not speak
How could they exchange love songs
During the tender nuances of togetherness
The lyric of her passion
The music of his muse

One falling summer evening
As the humidity held the air hostage
She and he, frayed in some places
Shared the refuge of a living-room couch
Frayed in some places

Sipping iced tea from perspiring tumblers
Welcoming the oscillations of a rusting window fan
Her cresting moon-eyes spoke volumes to him
He read every rising movement of her unspoken vernacular

Talking toward one another
As those in conventional dialogue seldom do
In the resonance of silence
Creating words no dictionary could house
Breaking the rules prescribed for the mute, the illiterate
Sensing and saying
All that is needed, all that is known

Pricks and Roses

Roses grow above pricks of green
Roses blossom; pricks are preened
Roses rise to kiss the sun
Pricks point fingers at the beautiful ones
Roses heal invisible scars
Pricks know precisely whom they mar
Roses inspire the ink of poetry
Pricks trigger the guns of graffiti

REVISIONS

If sorrow could set with the sun, the sun
Grave misfortunes at last be won, be won
Mounting mistakes subside to none, to none
If errors weren't echoes—redone, redone

HOLIDAY VIGNETTES
For my nieces and nephews

A perk in your step
A pinch in the wallet
Curled-up lips
Topped with a cherry nose

A flurry of snow
A scurry of shoppers
The celebration of a son
From whom this spirit arose

Evergreens in our homes
Wise men on frosted lawns
Earth-kings too devoted to Christ
To come in from the cold

A red bow on a beagle
A golden gesture from a loved one
The gifts of Our Savior
His story retold

Cookies coming to life
In the shapes of Santas and toy soldiers
Relations chatting and prattling on
Until their faces are blue

Child-angels in flight around me
The receiving from giving
My wish list is empty
My dream has come true

THE ABC'S DE ECSTASY
For Gilligan, A to Z

Any bad-ass can deal ecstasy
Faggots, gangsters, hookers, ignobles
Justice keepers laundering money
Needing only potential quick-fixes, random subjects
To usurp volatile wealth
Xanax,
You're zilch

THE FINAL TURN:

Exact your position
Less you diminish
Those still in condition
At least will finish

WHIPPERS AND TRAINERS

Know the difference, my lads,
    between whippers and trainers
Those who prompt you along,
    those who push you aside
Those you knead you carefully,
    those who need you incidentally
Those who preach in your face,
    those who teach beneath the surface

Note the difference, my lasses,
    between whippers and trainers
Those who mold your mind,
    those who scold your behind
Those who won't let you lay waste,
    those who want to lay you for a taste
Those insensitive to your yearnings,
    those sensitive to life's turnings

Know the difference, sons and daughters,
    between whippers and trainers
Those who help you hit your stride,
    those along for the ride
Those who see your potential,
    those who require your credentials
Those who circle when you are lame,
    those who go full-circle in this game

OZ HAS SPOKEN

Oz has spoken to your ruby shoes
--Head back home, or they'll be black and blue

Oz has spoken to your brain confused
--Honey, you only think you are being abused

Oz has spoken to your breaking heart
--That's not the only piece of your body I can tear apart

Oz has spoken to your once intrepid soul
--Swallow that courage, swallow it whole

Oz has spoken to your folks in their Kansas home
--It is all her problem and her fault alone

Oz has spoken to make one thing certain
--I am just a tiny man behind a curtain

THE MEADOW GROUNDS

The Meadow Grounds is a crying cloud today
Pony tails tumble in disarray
Ladies sport parasols but need umbrellas
Their gloves grab the overcoats of their fellas

The Meadow Grounds is a marsh today
The race is canceled; the annulled horses run astray
The fortunate saddles are able to dismount
Before the debacle swells and surmounts

PASSIONS OF POMPEII

In the 1860s
Preceded by a light dusting
Mt. Vesuvius erupted
Pummeling the city of Pompeii
With pumice and molten lava
Graying the sky, gloving the sun
Fallen figures
Hardened
In a cocoon of ash
The excavator Fiorelli wondered
What lay beneath the community of stone
He created casts and molds
Pouring curiosity into each cavity
To reveal Pompeians' love for art
A zeal no legal parchment could deny

In the 1990s
Preceded by light feuding
Our passions subsided
Pummeling our hearts
With remorse and rancor
Graying our souls, gloving our son
Fallen figures
Hardened
In a cocoon of ash
We wondered
What lay preserved beneath the common property of stone
We created forgiveness and hope
Pouring faith into each cavity
To reveal two persons' love for one another
A zeal no legal parchment could deny

Wilde-Ing

I wish I had been alive to toy
        with Oscar Wilde's indiscretions
A trade boy to torment him
Sartorially dapper, financially deprived
Destitute of morals
Even poorer in the pocket
I'd seduce him, enrage him, inspire him
To write the most chimerical, yet angst-ridden play
Or perhaps a tragedy in one act
The audience agog, the critics in awe
They applaud him, approve him, appease him
Before the conscience of the community
Imprisons him
Leaving me unshackled to walk and work the streets
A citizen making his mere contribution
To such deserving hands

Old Early

You became old early
Tired thoughts breathing faintly in his face
Your mind searching for good reason
His eyes scanning for any other place

Oh, you can still be clever
He can still be kind
Longing to leave gave him clear vision
Holding on to the untenable made you blind

I hear you are defending the defenseless
Standing by an absent man
The mother of invention and excuses
Rationalize it any way you can

Yes, you became old early
Not a wrinkle on your face
No need to polish your tarnished pedestal
Another trophy has taken your place

SONNET.COM

.my love is a virus of rare disease
.the diagnosis terminal and bleek
.all systems once go, aborted and weak
.sweet hour of prayer, I am on my knees
.i cast my dismay to a thousand seas
.where ships of fools float flawlessly then leak
.the moment of high tide has passed its peak
.my love is a walrus netted and dead
.albatross, fly over this corpse of stone
.i felt with my heart and ignored my head
.surrounded by silence, I die alone
.love's fate is unspoken yet not unsaid
.the soul is a warehouse for fat and bone

LOW NOON

Talk on, Storyteller
Disarm your pen in hand
Tell your tale
As your tongue paints a coat of many colors
My naked face becomes pale
Speak of an angel in the morning
As the sun goes up, her hungry desire goes down
Midday lunar lovers
Make ray-of-light love to their homebound prey
But those slick wolves never stick around

Another nervous day, another nervous breakdown
A daughter weeps to her mother of an unrequited love
Drenched by the deluge of an empty rain
No refuge for the homeless pain
The shaming sky exposes the unspoken truths from above
Talk on, drugstore novel queen
Speak of the seduction swaying through your shadows
Lovers make good liars
Liars make good writers
Perhaps you should put your pages of pain up for sale
A profitable sale

From Fiction to Folly

I saw the ghost of Mark Twain
On a riverboat casino
He was playing Twenty-One
Between hands, the hero from Hannibal
Told the dealer about a frog from Calaveras County
His throat croaking just for fun

Old Sam told the valet to repark his raft
And then go for a swim
He said the seat to his right
Was reserved for his buddy Ol' Huck Finn

When the white-haired word-wizard
Lost his last chip
Mr. Clemens begged for clemency
Upon refusal, he stood and stated proudly

I am glad this is one mess on the Mississippi
Tom Sawyer did not have to see

LYING, STILL

My husband loves our daughter
It is deeper than you think
He performs beyond paternal duties
His hands reach down around the pink

I have seen him slip her money
Witnessed her slip him the smallest smile
I never buy her clothes
She has all the latest styles

Our daughter loves her father
He is her only friend
My arms are folded neatly
I have no hand to lend

My neighbors voice their frustrations
For I am a sleeping slave
This mother would do something about the violation
But I am speaking from the grave

EQUINOX EVOLVING
For Mike

The fruit
Of days hanging
On the corner of Memory and Vine

The leaves
Of life falling
On a street reserved for sweeping and crime

The winds
Of fall rising
To beckon the hearkened waiting the call

The sighs
Of sun setting
As summer pulls down the shade on her wall

CARD GAMES

The dealer gave birth to aces, diamonds, hearts, and jacks
Her maternal instincts brought quite a deck of cards
        into the Casino
The pit boss forced her to separate this sibling stack
He commanded her to shuffle them by suit
Then place each set at four ends of the family card table
Play—he instructed—*play War!*
The high card wins each round;
and when the game is over, gather what you have and
        play again.

Eventually the dealer stood aside and watched him
        orchestrate these guerrilla games
Like a strategic Jersey shore mobster, a Vegas insider
He rejoiced in every victory; she lamented every defeat
After a few rounds, he decided who could play well and
        who could not
So he began to place wages and invest
        time, money, and pressure
The aces soon lost their edges
The diamonds depreciated in value
The hearts became broken, and the jacks became fools

After thousands of hands, everyone stopped keeping score
Everyone except the boss of this pit

The Silences Know

My love, what is to become of us
When our waking desire has gone to sleep
When words are worthless and talk is cheap
Only the silences know

Sweet solution, where do the answers lie
When our feelings move beyond mere touch
When our lips have moved too much
Only the silences know

Only the silences know
When our sense to stay
Segues to touch and go
Only the silences know

My dear director, how am I to react
When your familiar face looks strange
When the world offers nothing but change
Only the silences know

You approach with a persuasive tone
My belief in you still retreats
Somewhere between the cold heartbeats
Only the silences know

Only the silences know
When our tongues won't tell us so
When the hold around our hearts begins to let go
Only the silences know

My sound of joy, how will our song play on
When the music begins to disappear
When we call for tuning and no one hears
Only the silences know

The Silences Know

The pundits of pleasure agree
Lovers on the edge must communicate
But what to do when the advice arrives too late
Only silences know

Only the silences know
Who will fuel the flame
When our love-lantern burns low
Only the silences know

Mirabilia: Wonder of Lost Love
For Wynn, for always

How slender the breeze
Feather-dusting across the sand
How tender the tease
As an old lover took my hand
Wounded spirits, reminiscent souls
Lifting shadows from long ago
Sliding out, sliding in
The secret circles we know

Will we surface sunken memories
Netted in life's rolling sea
Will castaway hearts bank upon the shores of love
The soulful shores of love
Will the flora bloom again
From the bareness of these sequestered hands
A time for hope, a time for healing
The losses of this woman and this man

How voiceless the words spoken in such rash dismay
How joyless the birds as we soar our separate ways
An uncharted flight to a place unwanted and unknown
Riding out this tempest tossed,
Regretting our tempers having flown

Will we surface sunken memories…

A CIVILIZED MAN
For Doc Harper

My tongue is heat
I speak in flames
My breath circles like Dante's ringing burn
I seethe the caustic shame
My words infernal
My saliva, acid rain
So come, my sweet
Your mannered mouth must be untamed

My sex is heat
I fuck in flames
My manhood dips into the sauce
I seduce with no shame
My back is arched
My semen, acid rain
So cum, my sweet
Your chastity-lock must be unchained

A civilized man
Barbaric and ablaze
I wet my lips
With a lustful glaze
Woman, obsess on me
Throughout your days
I part your legs
With this parting phrase

CANTINA SESTINA

Each weekday at five, I head for The Cantina.
Benito Martinez is the bartender.
My drink of preference, of course, is the margarita.
I mind my manners and never get drunk.
My favorite thing is nibbling salt on the glass.
Someone usually plays the joint's jukebox.

It is 5:18; nothing is on the jukebox
as I stroll into my favorite cantina.
A guy shouts, *Hey, buddy, look out for the glass!*
I look toward the bar but see no bartender—
just Jeb Meyer on a stool; we call him The Drunk.
The love of his life is a good margarita.

Jeb tells me to play *Magarita-
ville*. It is D-16 on the jukebox.
(Now there's a surprise request from such a drunk.)
A neon light comes on beaming *The Cantina.*
Suddenly, I see the once absent bartender.
He's mumbling expletives about the broken glass.

To save a few bucks, I bring my refill glass
and order a Gold Primo Margarita.
Don't hold back on the Cuevo, my good bartender.
The Jimmy Buffet tune seeps from the jukebox.
I see the broken pieces on the cantina
floor. Jeb probably did it—who else but a drunk.

It is now 6:19, and I have drunk
a few Primos from my discounted glass.
The evening crowd spills into The Cantina—
too late to get a buck off a magarita.
The Happy Hour light dims above the jukebox.
A couple complains to the bartender.

CANTINA SESTINA

Don't whine to me; I'm just the bartender.
Benito retorts as his temper becomes drunk
with anger. He pounds the innocent jukebox.
It's been a long day—what with the broken glass.
I pity him as I sip my sixth margarita.
The neon now blurring the words *The Cantina.*

Ah-my sweet cantina, my sweeter margarita!
Benito sweeps the glass behind the jukebox.
Don't look at me, bartender, I'm not the drunk.

FRENCH LICK

You came home late from the Sorbonne
  with a new kiss on your tongue
The thesis on Renoir, you claimed,
  had to be redone
I could have sworn the paper
  was at the cut-and-paste stage
Or perhaps it is we
  who are at the cut-and-paste stage

Impressions rise with suspicion
  as I lie down with you
I recall the painter's preference
  for female French nudes
As you flicker then fold
  against the pink walls of my cheeks
I am aware of the slight changes
  to your mouth's old theme

And so this is a sad love song
  for us upon your return
Your silent lips tell me of a secret in Paris
  waiting your call
We relapse into revision
  as your sentiments make their debut
On American soil
  French kisses submitted for my review

HOLESOME

If I could calm the mental maelstrom
Scheming beyond these troubled shores
If the ravens circling in my mind would fly
Nevermore
If by the mere movement of my hand
I could bring forth bounty to this vacuous land

$$\begin{array}{c} \text{I would} \\ \text{D} \\ \text{I} \\ \text{G} \\ \text{Would} \\ \text{D} \\ \text{I} \\ \text{G} \\ \\ i \\ d \\ i \\ g \end{array}$$

To unearth all possibility

If I could halt the flood of anger
Pouring from these four chambers of rage
If I could read from an alternative script
Or at least turn to another page
If I could only forgive and forget
Instead of remembrance and regret
If God would only give me the gloves and the tools
To search for the man and bury the fool

HOLESOME

           I would
              D
              I
              G

           Would
              D
              I
              G

              i
              d
              i
              g

To unlock this captivity

If I were to find leaves of sage
I'd gather and take them home
I would sip them in my tea
Until I could discern from loneliness and being alone
I'd drink from this cup of wisdom
Take the knowledge in
The education steaming in my primer-soul
To ignite the lightless pupil from within

           I would
              D
              I
              G

           Would
              D
              I
              G

              i
              d
              i
              g

To unleash this fantastic reality

CHURCH MOUSE

A church mouse stands in his blessed corner
Dormant to the whirlwind outside this hallowed plywood
--Socialized at birth
        --Institutionalized for life
A reticent bishop regaling in his privilege
       gnawing on crumbs of cake
Residue from last evening's marriage of like-minds

It is Sunday morning, his night to howl
He greets the members of his cloth
As they herd through the alcove like cattle
Hung over from Saturday night's pasture grazing
Hoofed in the meat market of their hypocrisy
Tripping into pews
Mooing and mumbling the invocation
The designated drivers mouthing the words
Punch-card Christians savoring the holes
In their tickets to heaven
A choir of liars cognitively concentrating
On the upcoming pigskin games this afternoon
Withholding a tithe of their offering to wage
On the Saints of Sin City
A wager of piety; a wager of frivolity

These secular confessions fail to escape
The homebound church mouse
A shoreless sandpiper sowing his oats
Black, tan…well, you know the rest
His judgments a swelling edema of holy water
Bloating his body, drowning his brain
Anchoring his appendages
Unable to reach out
Unable to reach in

Nettlesome
Intractable
Leering
Disdainful
Austere

CHURCH MOUSE

This is Eden, and the revered rodent has bet
His shriveled mammary glands on it
His steeple's way is the only way
It points straight up
Like an index if you are one of them
Like a middle finger if you are not
Indeed—or, at least in deed, his way is the only way
And there is no way out

WILD DESERT ROSE

The heartland awaits to reclaim you,
My wild desert rose
Your grace is like a garden in full bloom
Bless you for your buds of buoyancy
As my undulating confidence came undone
You withheld me from wayward ruin

My garden desires you, wild desert rose
The soul cultivates no more natural thing
Another nightfall needles me
As I lay myself on the longing for your loving
Sheets of solitaire, linens unfolding for one

Save the last leaf and thorn for me, wild desert rose
When your burgeoning beauty
among the sandswept dunes is done
Oh, let the loss of first love at last be won
Wild desert rose

A fool never to have told you
When the privilege was mine to hold you
How I only wanted to unfold you
And I do
I do

THE FINISH LINE:

Roses for celerity
Woes for the slow
A moment of celebrity
If you win, place, or show

BREAKER

To pasture or to glue
As if the choice were up to you
A stallion to stud
Slip out of your genes
Share some blood

Night mare or evening sire
Where do you roam when you retire

Mama's little stable boy
Late to the gate, the last to start
To break from the field
You must break some hearts

FLYING DAYS

My flying days are falling
The winded hour is through
Time to stay more grounded
And wear sensible shoes

My sailing days are sinking
The sea-breeze kiss bids adieu
Time to perch among the sand's shells
And watch the sunset view

My riding days are retiring
The stallion's strut is through
Time to put the breeder to clover
Before he turns to glue

My dancing days are unwinding
The four-footed waltz is down to two
Time to plant myself on the front porch
And stabilize these roots

My breathing days are bowing
The on-the-air show is through
Time to thank you, folks, for coming
Bless you for staying tuned

DEAD TO THE WORLD AND WIDE AWAKE

I am dead to the world and wide awake
Born for hell's purpose, for heaven's sake
I am a fragile boy breaking with no glass
I will reveal the content of my hidden package to you
If you only ask
I tell a sad story
With every twist and hook
Get out your handkerchiefs
Open your reference books
Like a bad meal, the tale often repeats
When the fingers of a trusted hand probe between the sheets

He whispered, *Surrender, and I will love you tender*
I whimpered, *But I am not dreaming anymore*

Caretaker, if I were to exhume our secret past
While you closed your eyes and counted to ten
Would you let me run and hide far enough from you
To seek asylum from our bedroom Bedlam
My devil of delight, my angel of pain
You stole my sunlight, and replaced it with rain
A feathered case in my face
To muffle the latest attack
I begin to dream in black and white
The colors of childhood fade away
They won't be coming back

If a course of action were clear, I would redress
If anyone were to lend an ear, I would confess
Oh, if I could, I'd tell the neighborhood
Or perhaps I should just let   it   go

He whispered, *Surrender, and I will love you tender*
I whimpered, *But I am not dreaming anymore*

Living in a glass bowl at home
I became a flying fin in a public school of fish
How I splashed and learned
When class was dismissed
I crashed and burned

DEAD TO THE WORLD AND WIDE AWAKE

My devil of delight, my angel of pain
You stole my sunlight and replaced it with rain
Well, big boy, you've got yours
Now I've got mine
Meet me for the final showdown
At our hometown five-and-dime

I am dead to the world and wide awake

PHILOSOPHICAL RESIDUE

Oh, Freezing Flame
Spare the philosophical poets
Who else will speak of earth's end
At earth's end
Who else will afford the mutual respect to listen

Ice and fire are kissing cousins
Two kids on summer vacation
Engineering homemade bombs
For the winter solstice

This eve of new year never
--no time for oiled paintings
--no time for group pictures
Faces at monitors
Keyboarding farewells
E-mailing exits, logging out
The big log

Oh, spare the dramatic poets—at least revive Eliot
To comment as we become shards of ash-frost
His eloquent elocution during our massive execution
Terminal

STRUNG INSTRUMENT

A marionette

D
A
N
G
L
I
N
G

On angel-harp strings

H
A
N
G
I
N
G

From heaven

Each moment rehearsed, each word scripted
His lifelines wrapped around the finger of his God,
His Gepetto
An actor upon the stage of life
Until the drapes of death descend
The encores echo to silent vibrations
Heard only by night-winged creatures
Who fly through the strings
Detatching him from darkness
Lifting him home to the stone, home to the light
His performance critiqued upon paper
Sealed with approval
He enters the unknown
With no strings attached

Tolled Life

I, being fated a man, must confess
My muscled arm and testosteroned brain
Are merely tools upon God's broader plain
Not unlike a pearled neck or wind-blown dress
This high truth makes me no more or no less
A person than the lofty born again
Their appendages anointed with rain
From sacred skies where tears of angels bless
I, being a hated man, take a stroll
Down crowded streets with nary a soul there
Faces judge me whole, though see me in part
I come to the Gate and inquire the toll
Humanity's price is an unfair fare
I offer coins, but the price is my heart

First and Last Date

My date Lynnette looked at my bowl of half-eaten
        rice casserole and quipped,

There are people starving in China, you know.

I glanced at her half-empty bottle of lukewarm Corona
        and said,

There are people sober in Mexico, you know.

BREAD AND CIRCUS
For Gene

Star-crossed lovers confuse a dream with despair
All in love is unequal, and the imbalance fair
Baez sang to Dylan of diamonds and rust
Deserting her was unjustified; and his desserts, just

Tristan brought Iseult to Britain to be a bride to King Mark
The two commoners fell in love; and their fate became dark

Now, astral voyager,
Ignore the song of the siren, wrap the reins tightly around
          the Winged Horse
Venture not from your known environs
Maintain good sense and good course

For the charade of love is circus and bread
Folly and free food to delude a clear head
Bread and circus is the more accurate phrase
To describe wool over the eyes, the drape of a haze

The Romans did this to raise an empire
Impatient lovers use the technique to ignite a false fire

The mere passing of eyes unlocks the temptation
The mere passing of time proves the condemnation

THE ANTI-PARADISE
For Dr. Shields

In the South Pacific Ocean
East of Bora Bora
North of Puka Puka
Floats Disappointment Island
I have been there without being there

The place belongs to no one
No country bothers to fight for its possession
Though the isle possesses the citizens of all nations
(And some refer to America as the sleeping giant)

The main export is loneliness
The mound of tears imports lost souls
A passport is not required
All sorrowed sojourners are welcome

In the South Pacific Ocean
East of Bora Bora
North of Puka Puka
Floats Disappointment Island
Where scars are souvenirs

For more information
See your travel agent

Or build a house of hope

Or dance in a room of dreams

EVERYBODY'S ANGEL

The unchosen were lost until I came...

I am everybody's angel
Admire the halo upon my head
Guaranteed a rite heaven
I cannot wait until I am dead

A precious little angel
God's own sweet freedom dove
You need a dose of my religion
A taste of my saccharine love

The unchosen were lost until I came...

Everybody's angel
You must place me in your trust
I will pummel you to powder
Then I sniff the angel dust

You are drowning in sin and sadness
Destined to a sea of tears
What I earned merely by birthright
You will strive for in vain for years

The unchosen were lost until I came...
The sky was crying until I came

If I am accused by a thousand eyes
I will be relieved, I will be relieved
For I am justified
In what I believe, in what I believe

Everybody's angel
My shit-eating grin seems so true
Fall at my feet, then kiss my heels
As I turn my back on you

Such a precious little angel
You need not worry about a thing
Lay your breast underneath my bosom
Wrap your legs around my wings

WINTER: AN 8 BY 8 PORTRAIT

The whirling snow a carousel
Of winter amusement and joy
Flakes falling to blanket, comfort
The exhausted lay of autumn
A promise of April beneath
December's carnival of white
Life's crystal anniversary
Winds shake the bloom of frost's flower

EARNING MARKS

Chalk marks of thought upon an empty slate
A dark field of blanks ignited by fire
Each fissure unfolds to anticipate
The stimulation of learning's desire

The mind is a canvass in need of paint
A portrait of space hanging in a hall
Brush lightly, Maestro, or else you will taint
The rise of pure Genius along these walls

HORSEMEN FROM HEAVEN

When man breaks his last promise, the Lamb breaks the seal
The Apocalypse comes riding on horsemen four
Led by *War* mounted upon a stallion of white
To *his* left walks Conquest, to the right of *him*, Christ

Slaughter swiftly follows with peace on earth to steal
Violence accompanies *him* to settle the score
A mighty sword brandished above a steed of red
All the living who pass by *him*, stricken and dead

Famine's tertiary role takes mankind's last meal
Corpses and crumbs fall and pile to reach Peter's floor
Pestilence and Poverty guide the black creature
The absence of substance is *its* finest feature

Death travels alone with no assistant at heel
He reviews the work of the three coming before
His scythe sharpens the sun; *his* mount nameless and pale
Without uttering a sound, *he* bids all farewell

Best be wary of horsemen who ride in quartets
The Lord is forgiving, but He never forgets

BABES IN BASINS
For Helen and the Hampton Family

When her children were young, she bathed them in tiny basins, cleansing their concerns, anointing the scrapes of another day at play—water from her palm pouring onto their sensitive skins, the touch of her hand a visionary guide prompting them to bed and the dreams of beyond.

As her four tadpoles outgrew the confines of these bubbly bowls, she showed them the rivers where they could wash on their own. Helen threw out the bathwater but never withdrew her love. The children always knew they could swim back home.

After their mother heard she was dying, the woman gathered new basins, though her children were cleaning kids of their own. She filled the baby pools by clipping and collecting moments from their days of wonder and wander to nurture their memories when she was gone.

A photograph lasts longer than the splash of warm water; a fraying yo-yo returns to the grasp of an awaiting finger-fold of old. Helen left no long letters to explain her departure. The sea of stuff filling the basins was the story she told.

THE HAZE OF DISASTER
For John F. Kennedy, Jr.

The clouds like chefs in white, billowy hats
Prepare and serve a storm
Yet I am unable to smell the first drops of rain

The aroma of loss is everywhere
In the salts of the sea, in the salt-tears of crestfallen faces
The haze of disaster
Osmotic as it penetrates
The walls of Camelot
The ironclad walls of Camelot
Where royalty gathers to eat the feast
Another day of ambrosia in Camelot
The fruit of labor, the raw meat of money
Slabbed on the tables of Camelot
The fumes of fortune rise, then subside
To the scent of disaster in Camelot
The foul air spoils the fruit and taints the meat
Behind the guilded walls of Camelot
The fragrant fall of power to what the gods have wrought

BETWEEN GOOD FRIDAY AND EASTER SUNDAY

I awakened in the past again
All around me the future was forming
Iridescent iris opened the eye of the sky
As night spilled into morning
A thunderous thought circled my cerebrum
With no storm warning

The cold rush of rusted bathroom-sink water
Set my slumbering mind wandering
As I prayed to reach beyond the point of pondering
The sooty sins of my soul
In desperate need of laundering

What will a perusal of the Scriptures teach me
That the sanctified and their sermons have not depicted
If these indeed are the lines of the Lord
Why must they be black-and-white scripted
Oh, how I long for the tenant of my turmoil to be evicted
The Angel of Mercy appeared before me and proclaimed

Wayward Son, get this straight
 before your own body is crypted
A thousand and a thousand years ago
Your Savior was left there swinging
A death knell heard around the world with no bell ringing
No incense, gold, or myrrh could surpass
The gift Jesus himself was bringing
As He rose from the dead
The sins you were yet to commit
Forgiven

As the Winged Wisdom took flight
My enlightened eyes sunk in dismay
For I had forgotten to inquire about the problem
Probing my brain this one day

What was Jesus Christ doing
Between Good Friday and Easter Sunday

CENTURY FALLING
For Inez Newquist

One-hundred years, one-hundred letters
Blue rivulets run through my age-worn appendages
As the blue ink in my dowry quill runs dry
I regret not to have loved you better
The wonder of unspoken splendor
Palpable pleasures never die
A century is falling from the sky

My life maintained by an HMO-approved agenda
Supervised by a licensed badge with a master's degree
Nurses and attendants
The credentialed bear no resemblance
To the care your comfort offered me
So graciously

The social souls meet every Thursday
Card games and bingo for a little levity
Each play, I pray no hand draws the ball reading O-63
The month, the year my Texan had to leave me
The pain persists, Dear Edward, believe me
Though etiquette suggests a lady seldom cry
As her century is falling from the sky

So this, my final and most forthright letter
Affixed with no paper-glue portrait of the Queen or Elvis
To catch your omniscient eye
I beg to differ with Sir Isaac Newton
The fabled fowl most justified in her hollering and hooting
A century is falling from the sky
Our millennium draws nigh

The Scent of Your Ghost

Some evenings, Louise, when the day has been full
And the blazon-moon ignites my eager memory
I carry my undying flame for you through the rooms
Where I must remain living

With the lights down and the sounds of living subsided
I wait for the scent of your ghost—
A scent evoking more than smell
An aroma haunting my senses with traces of you

The scent of your ghost is the whisper
Of apron strings binding around your waist
As your skirts and scarves lie in a box
For the Salvation Army

The scent of your ghost is the rustling of letters
Upon the desk in the alcove,
Though I now receive all bits of mail
At the office

The scent of your ghost
Sweeps October leaves from the back doorstep
As the Norman Rockwell calendar patters
A portrait of April rain

The scent of your ghost simmers sauces of oregano and basil
Upon a barren stove in an abandoned kitchen
When the meal no one will eat tonight is finished
A hint of perfume on your gossamer dress calls me upstairs

As I rise to meet the risen
I discover no foot stands at the stair landing
I fold my hands, scoop the air before me into my arms
And carry the scent of your ghost, my love, to bed

The Going Rate

What is the going rate
For keeping sane
Playing it straight
Having a choice
Peanut or plain

What is the going rate
For standing still
Longing to wait
Having a voice
Silence to kill

What is the going rate
For shunning pride
Leaning toward fate
Showing yourself
Nothing to hide

What is the going rate
For coming home
Cleaning the slate
Being yourself
One of our own